AF797565

# Organization Is the Key

1 O 16.00 | 2 r 17.50 | 3 g | 4 a 18.65 | 5 n 19.16 | 6 i 19.82 | 7 z 22.03 | 8 a 25.44 | 9 t 26.01 | 10 i 26.86 | 11 o 26.92 | 12 n 28.70

Harcourt
SCHOOL PUBLISHERS

Orlando Austin New York San Diego Toronto London

Visit *The Learning Site!*
**www.harcourtschool.com**

## The First Atomic View of Life

More than 2,000 years ago, a Greek teacher named Leucippus made an observation to his students. He showed them that if they looked out at a beach, the sand looked smooth and continuous, like the ocean. But, if they held up a handful of sand and looked at it closely, they would see that it was made up of tiny, indivisible grains, or particles. Leucippus was making this observation to help answer a question that many people of ancient Greece had tried to answer: What is the world made of?

**The four elements**

One student, Democritus, developed his teacher's observation into a theory. Democritus believed that all matter in the world was made up of tiny particles. The particles were so small that they were invisible as well as indivisible. He gave these particles the name *atomos*, which in Greek means "cannot be cut." (You probably can recognize where the word atom comes from.)

Leucippus and Democritus were ahead of their time, and their "atomist view" of the way that all matter in the world was made up would not be recognized for centuries. However, their ideas would be among the first steps on the path that would eventually lead to the development of the periodic table of elements.

## The Four Elements

Although Leucippus and Democritus believed that all matter was made up of tiny particles, most Greek thinkers in ancient times believed that all matter was made up of four basic elements: earth, air, fire, and water.

They also believed that the elements existed in combinations of hot, cold, wet, and dry. These became known as the classical elements and were usually listed in the following way:

Fire is both hot and dry.
Air is both hot and wet.
Water is both cold and wet.
Earth is both cold and dry.

The Greek philosopher Aristotle popularized this theory of matter. He believed that everything in the world was made up of combinations of earth, water, air, and fire. Each of these elements could also be transformed, or changed, into another. Greeks in ancient times believed, for example, that water could be changed into air by boiling it. For centuries, Aristotle's ideas had a major influence on the way people viewed the world.

## Discovering the Elements

Aristotle, as well as the other Greek thinkers of his time, developed their theories of matter through reason. By the 1600s, scientists would begin to use experimentation to test their theories.

Elements such as gold, silver, and copper have been known to people since ancient times. Weapons, jewelry, and everyday household items have been fashioned from these elements for thousands of years.

In 1669, the first scientific discovery of an element was made—somewhat by accident. Hennig Brand was a German chemist who was looking for a way to improve his fortunes. During this time, some people believed that matter could be transformed into something else. Chemists of the day often tried to turn ordinary metals into gold. One scientist was even paid by the Dutch government to try to turn its beaches' sand into gold.

Brand tried using an unusual substance, human urine, in his experiment. He boiled it down to a paste and then heated the paste, hoping that it would condense down to gold. Instead, he was left with a white substance that glowed in the dark!

This substance would come to be known as phosphorus, which means "bringer of light" in Greek. Brand did not make gold, but he did make an important discovery. He tried to keep his discovery secret, but an English scientist named Robert Boyle saw it exhibited and, in 1680, came up with an improved method of producing phosphorus.

Robert Boyle is important to the story of the development of the periodic table because he challenged the idea that all things are made of four elements. He thought that this theory could not be proved through experimentation. He said, "Out of some bodies, four elements cannot be extracted. . . ." He believed that elements could be found only through experimentation, as Brand's accidental experiment showed.

## Scheele and Priestley

About 100 years after Brand discovered phosphorus, a scientist in England, Joseph Priestley, and an apothecary in Sweden, Karl Scheele, were doing experiments that would help lead to the development of the periodic table.

Robert Boyle

Karl Scheele

Joseph Priestley

Priestley's discovery was the result of an accident, just as Brand's discovery had been. Priestley had mixed two gases obtained from heating a mercury compound and red lead. He let them stand overnight, thinking that in the morning the mixture would be completely unable to maintain life or a flame.

Although he had no plan to test the "air" with a flame, he impulsively thrust a burning candle into the jar the next morning. He was astonished to see the candle burning quite brightly! Further tests proved his suspicions that the "air" from burning mercury was very different from any that he had tested so far. He published a paper announcing that he had discovered a "new species of air."

In 1772, Karl Scheele was working for an apothecary in Sweden. He spent his days mixing medicines for his employer. He spent his nights, however, experimenting in the back room of the shop.

Scheele was curious about the true nature of air. He called all gases "air" because he thought of them as "air elements." He spent weeks collecting the "air" given off from burning phosphorus, which was used in that time to make medicines. Saltpeter (potassium nitrate) was another substance that he often burned. Scheele poured the saltpeter into the glass tube and placed it over a small fire. As the saltpeter burned, the gas collected in an empty beef bladder.

To test each gas, Scheele alternately placed mice or nearly extinguished embers in glass containers filled with each gas. In some of the jars, mice died; in other jars they thrived. In some jars, burning candles went out instantly. Scheele wrote down all of these results and wondered what they meant.

One day, Scheele burst into the shop waving an "empty" bottle stopped with a cork. He was shouting "Combustible air! Combustible air!" Then, to the astonishment of the customer and his employer, he lit a splinter of wood, blew it out, and dropped it into the bottle. It instantly burst into flame again! They thought it was magic, but Scheele knew that he had discovered a new kind of "air" that was vital to fire. He called it "fire air."

Priestley and Scheele had actually discovered oxygen, although that was not the name they gave it. That name would come from someone else.

## Antoine Lavoisier

Antoine Lavoisier had a great fondness for science and all things related to it. In fact, he has been called the "father of modern chemistry." By profession, however, he wasn't even a scientist. He worked for the French government and in private business in the 1700s. At night, he and his wife, Marie-Anne Paulze, conducted experiments in the lab that he built.

Chemists from all over Europe came to the lab to learn from Lavoisier. Marie-Anne learned to speak English so that she could translate the works of famous British scientists for her husband to read. In fact, she translated Joseph Priestley's work. She also studied art and engraving so that she could make meticulous drawings of her husband's experiments.

**The Lavoisiers**

By this time, scientists had discovered other elements in addition to phosphorus, but they did not have a systematic way to name or classify these elements. Lavoisier believed that chemistry needed to have its own vocabulary. At the time, one element was called "fixed air." In Lavoisier's new system, this element would be called carbonic acid. Salts that were formed by this acid would be called carbonates. For the first time, compounds were named for their component elements. Except for a few scientists who were reluctant to accept change, Lavoisier's work quickly became the standard and is still the basis of the current chemistry vocabulary.

Lavoisier was known for the hard work he put into his experiments. Like Scheele and Priestley, he burned different substances and collected the "air" they gave off. The name that he gave the air, however, was oxygen.

In 1789, Lavoisier published *Treatise on the Elements of Chemistry*. In this document, he listed the 33 then-known elements, using his new naming system. He cautioned that some of these elements may in fact be compounds. It would be left to other scientists to prove this theory to be true.

This was a troubled time in France's history, however. Because of his involvement with the previous French government, Lavoisier was executed during the Revolution. Marie-Anne would preserve her husband's notebooks and laboratory equipment. In 1803, she published his memoirs.

## Humphry Davy

Lavoisier's work reformed the study of chemistry. One of the first scientists in the new field was Humphry Davy. He showed no promise as a scientist at all while he was growing up. Until he was 16, he ignored his schoolwork. His teacher said, "He'll never set the world on fire!"

After Davy's father died, he went to work for a doctor. There, he began to be interested in chemistry and learned that he had a gift for asking the right questions and setting up useful experiments.

His enthusiasm and successes in chemistry became so well known that in 1801, the Royal Scientific Institution in London invited him to work for them. The Royal Scientific Institution had rooms full of equipment, so Davy went eagerly to London.

Electricity was new at the time. Batteries had just been developed a few years before, and the Institution had all the necessary materials to make more. Davy explored electrochemistry more than any person before him. He was hoping to discover new elements by breaking down caustic potash (potassium hydroxide) with electricity.

One morning after many unsuccessful trials, Davy decided to hold a piece of potash in the air briefly. He thought that if it picked up a little moisture from the air, it might be a better conductor for electricity. The potash immediately began to react with the air, forming a thin layer of water on its surface. Then Davy placed it on a sheet of platinum that was connected to the negative end of the battery. Holding a wire on the potash, Davy closed the circuit, and the current passed through the potash. It began to separate into tiny, shiny globs.

**Humphry Davy and his battery**

Davy named the shiny pellets *potassium,* but he was not sure what kind of substance it was. He noticed its properties. It was shiny like a metal, and it was a good conductor of electricity, but it was very light. Davy had always thought that metals were heavy. Then he remembered that iron was light compared with platinum or gold. So Davy decided that potassium was simply a very light metal. His discovery astonished other scientists, who had always thought that caustic potash was an element. Davy proved that it was made up of elements, one of which was potassium.

## The 1860s

After Davy's discovery, chemists around the world raced to find new elements. By the 1860s, chemistry was a jumbled mass of discoveries that seemed to have no order. Chemists all over the world tried to organize the new elements into some understandable diagram.

During the decade of the 1860s, several European scientists, working independently, came up with their own ways to arrange the elements. In 1862, A. E. Beguyer de Chancourtois, a French geologist, listed the elements around a cylinder. In this way, sets of similar elements lined up vertically. Because he was a geologist, he didn't use a chemical vocabulary, so his work did not receive much attention. He is now remembered, however, as being the first scientist to notice the *periodicity of the elements,* meaning that when the elements were grouped by their atomic weights, similar elements occurred at regular intervals, or periods.

The next year, an English chemist named John Newlands noticed a pattern when he grouped the 56 elements that were known at that time. He grouped the elements into sets of seven. He noticed that the first elements in each group were similar, as were the second elements in each group. In fact, this pattern held true for all seven elements in each group.

Other chemists criticized Newland's work. His contribution to the field of chemistry would not be recognized until 1887, when he received the Davy Medal for chemistry.

## Dmitri Mendeleev

One Friday morning in 1869, a Russian chemistry professor named Mendeleev was writing a chemistry textbook. He was trying to write an explanation of the order of the elements because he thought his book would not be complete without it. Mendeleev was working to meet a deadline—he had to finish this part of the book by the next Monday. Besides being a chemistry professor, Mendeleev was also a trusted scientific advisor to the Russian government on topics ranging from coal mining to cheese making. On Monday, he had to catch a train for another city so that he could give a speech to some cheesemakers.

Mendeleev did not sleep that weekend. He wrote the then-known elements on cards, along with their atomic weights. He searched for a pattern that he thought might be like the one in a card game he used to play called Patience. In that game, cards were laid out in a pattern going across and down.

Monday morning came, and Mendeleev still could not find what he was looking for. He dismissed the sleigh driver who had come to take him to the train. He continued to work through the day.

**Dmitri Mendeleev**

As it grew dark that evening, he was overcome with exhaustion and fell asleep at his desk. When he awoke, he recorded a pattern that he had seen in a dream. The arrangement followed two principles: elements arranged by atomic weight in vertical columns and elements in rows that have similar properties. When the rows were arranged on top of each other, the seven columns formed orderly groups of elements. Mendeleev had discovered periodic law—that an element's properties were related to its atomic weight. But no one believed him.

The trouble with Mendeleev's table was that it had blank spaces. His response to the science community's objections was that there were more elements to be discovered. He predicted elements with their precise atomic weights and characteristics to go into each blank space. Other scientists mocked his table and periodic law.

**Mendeleev's card game**

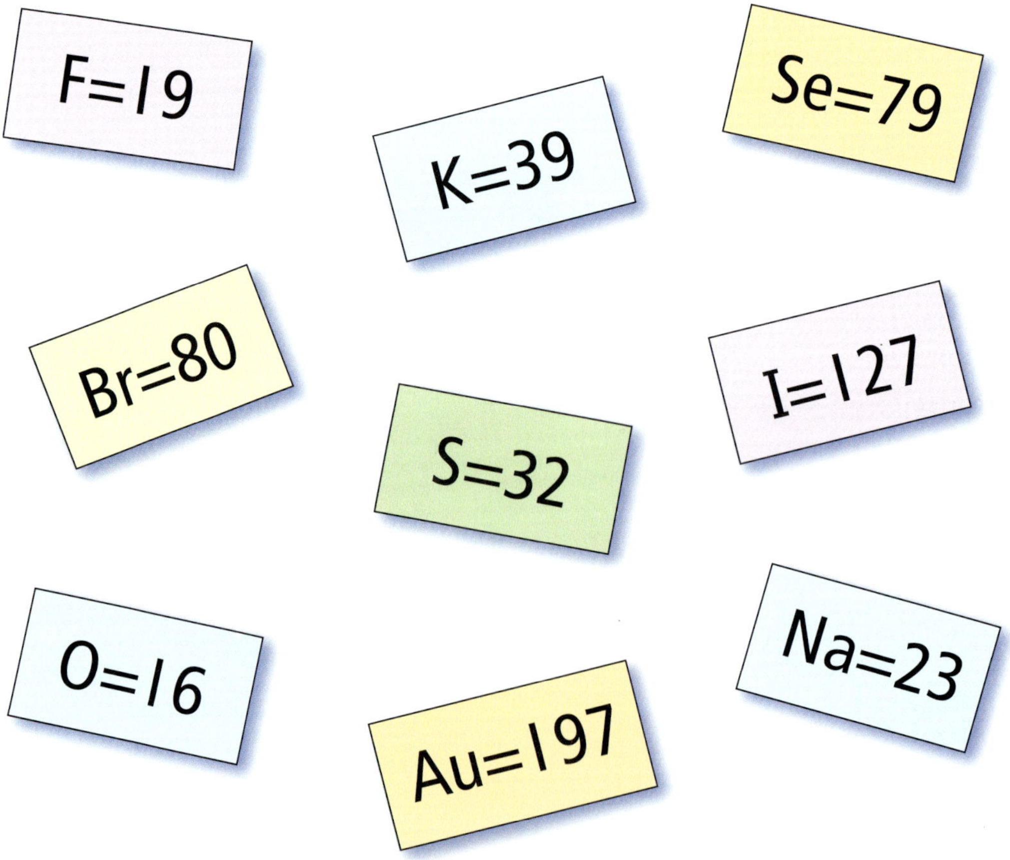

## Is the Table Stable?

As arguments about Mendeleev's table continued, chemists around the world raced to discover new elements and achieve fame. In 1875, a French chemist announced that he had discovered a new element that was very much like the familiar aluminum. The announcement, made through the French Academy of Science, included the weight of the element as well as other measurements of it that he took. Mendeleev quickly wrote a letter to the Academy. He disagreed with one of the measurements, claiming that according to his prediction, it had to be wrong.

### Time line

400 B.C. Democritus proposes that matter is made up of particles.

500-300 B.C. Greeks generally believe that all things are made up of four elements.

B.C.

1600 A.D.

1649 A.D. Henig Brand discovers phosphorus.

1661 A.D. Robert Boyle questions the four elements.

1700 A.D.

1772 A.D. Karl Scheele discovers "Fire Air."

1774 A.D. Joseph Priestly discovers "new species of air."

1789 A.D. Antoine Lavoisier publishes chemistry textbook.

1800 A.D.

1807 A.D. Humphry Davy discovers potassium.

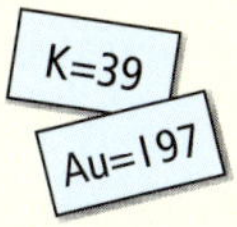


1808 A.D. John Dalton popularizes ancient Greek theory of atoms.

1862 A.D. A.E. Beguyer de Chancourtis groups elements on a cylinder.

1863 A.D. John Newfand classifies 56 elements.

1869 A.D. Mendeleev publishes the table of the elements.

1870 A.D. Lothar Meyer publishes similar table of the elements.

The French chemist scrambled to get more of the element and to test it for purity. Conducting new tests, with a purer sample, he admitted that Mendeleev was right. The element, gallium, belonged two spaces below aluminum on the table! It even had the properties that Mendeleev had predicted for that element. The world was slowly moving into acceptance of the periodic table.

The final proof of Mendeleev's law came in the 1890s in England. William Ramsay was analyzing air to identify impurities that he thought were present all around us. Following the usual twists and turns of scientific experiments, he finally realized that he had discovered helium and argon. The problem was that there was no place for them on the table. It seemed that the periodic law was not a law after all.

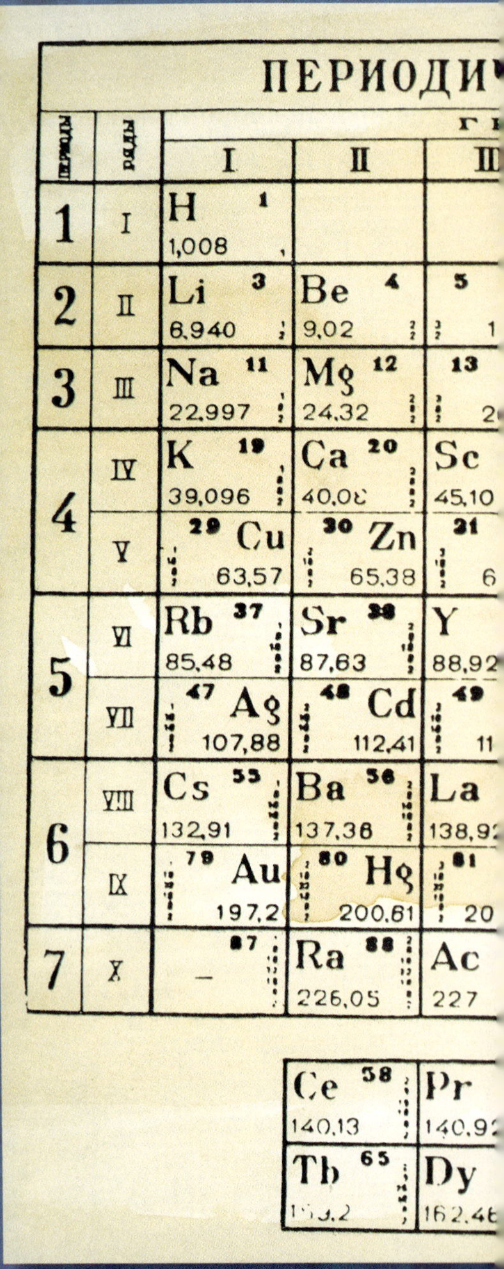

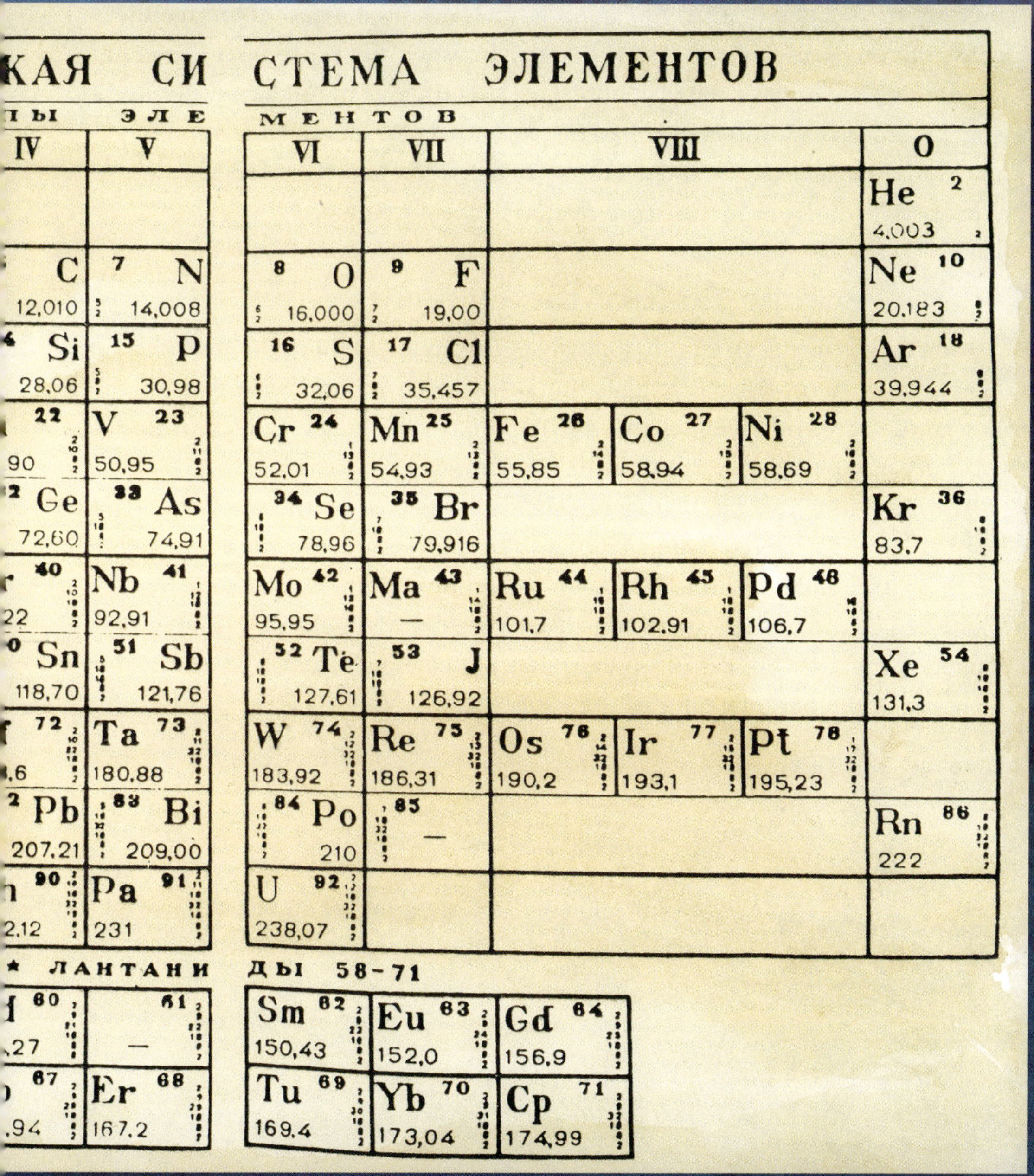

…КАЯ СИ СТЕМА ЭЛЕМЕНТОВ

ПЫ ЭЛЕ МЕНТОВ

| IV | V | VI | VII | VIII | | | O |
|---|---|---|---|---|---|---|---|
| | | | | | | | He 2 4,003 |
| C 12,010 | 7 N 14,008 | 8 O 16,000 | 9 F 19,00 | | | | Ne 10 20,183 |
| Si 28,06 | 15 P 30,98 | 16 S 32,06 | 17 Cl 35,457 | | | | Ar 18 39,944 |
| 22 90 | V 23 50,95 | Cr 24 52,01 | Mn 25 54,93 | Fe 26 55,85 | Co 27 58,94 | Ni 28 58,69 | |
| Ge 72,60 | 33 As 74,91 | 34 Se 78,96 | 35 Br 79,916 | | | | Kr 36 83,7 |
| 40 22 | Nb 41 92,91 | Mo 42 95,95 | Ma 43 — | Ru 44 101,7 | Rh 45 102,91 | Pd 46 106,7 | |
| Sn 118,70 | 51 Sb 121,76 | 52 Te 127,61 | 53 J 126,92 | | | | Xe 54 131,3 |
| 72 ,6 | Ta 73 180,88 | W 74 183,92 | Re 75 186,31 | Os 76 190,2 | Ir 77 193,1 | Pt 78 195,23 | |
| Pb 207,21 | 83 Bi 209,00 | 84 Po 210 | 85 — | | | | Rn 86 222 |
| 90 2,12 | Pa 91 231 | U 92 238,07 | | | | | |

* ЛАНТАНИ ДЫ 58-71

| | | | | |
|---|---|---|---|---|
| 60 ,27 | 61 — | Sm 62 150,43 | Eu 63 152,0 | Gd 64 156,9 |
| 67 ,94 | Er 68 167,2 | Tu 69 169,4 | Yb 70 173,04 | Cp 71 174,99 |

**The first periodic table**

Chemists around the world weighed in on the issue, but it was Ramsay himself who finally settled the question. He argued in favor of the table's structure. He decided that helium and argon represented a new column—the eighth column—and that the rest of that group was likely yet to be discovered.

Ramsay gave himself the task of discovering the rest of the group. By 1898, along with Morris W. Travers, he had discovered xenon, neon, and krypton. The world no longer doubted the validity of the table. It was a fixed law that the properties of the elements were related to their weights.

## Today's Periodic Table of the Elements

During the twentieth century, chemists continued to fill in the periodic table of the elements. Most scientists think that it is complete. Each one followed the pattern of Mendeleev's prediction. His mathematical genius saw the beauty of the natural pattern in the elements.

The only thing that has changed is the ordering rule for the table. Chemists have discovered that an element's properties are related more directly to its number of protons than to its atomic weight. Now the elements are ordered by the number of protons. Nevertheless, Mendeleev's idea that elements' properties appear in a pattern has not changed. He has remained the true father of the periodic table of the elements.